Farm Bee-Keeping
Farmers' Bulletin 138

by US Dept. of Agriculture

with an introduction by Jackson Chambers

This work contains material that was originally published in 1915.

This publication is within the Public Domain.

This edition is reprinted for educational purposes and in accordance with all applicable Federal Laws.

Introduction Copyright 2018 by Jackson Chambers

IMPORTANT NOTE & DISCLAIMER

IMPORTANT NOTE :
As with all reprinted books of this age that are intended to perfectly reproduce the original edition, considerable pains and effort had to be undertaken to correct fading and sometimes outright damage to existing proofs of this title.

At times, this task can be quite monumental, requiring an almost total rebuilding of some pages from digital proofs of multiple copies. Despite this, imperfections still sometimes exist in the final proof and may detract slightly from the visual appearance of the text.

Some images may suffer from reduced quality due to anomalies in the original scan.

DISCLAIMER :
Due to the age of this book, some methods or practices may have been deemed unsafe or unacceptable in the interim years. In utilizing the information herein, you do so at your own risk.

We republish antiquarian books with no judgment or revisionism, solely for their historical and cultural importance, and for educational purposes.

Self Reliance Books

Get more historic titles on animal and stock breeding, gardening and old fashioned skills by visiting us at:

http://selfreliancebooks.blogspot.com/

Introduction

Here at **Self-Reliance Books** we are dedicated to bringing you the best in *dusty-old-book-knowledge* to help you in your quest for self-sufficiency and food independence.

This special edition of **Farm Bee-Keeping** was written by the *Missouri Agricultural Experiment Station*, part of the *U.S. Department of Agriculture*, and first published in 1915, making it over a century old. It is also known as **Farmers' Bulletin 138.**

Farm Bee-Keeping has sections on *Value of Bees, The Life and Habits of the Bee Colony, How to Get a Start, Necessary Equipment, How to Handle Bees, Increasing Colonies, Honey and By-Products*, among others.

The book also features recipes for using your home-produced honey as an ingredient, and notes on Bee-Keeping in all four seasons of the year.

A super-short, fast read, and a great place to start for all those considering taking up Apiculture, or for those starting out in the field.

~ *Roger Chambers*

State of Jefferson, March 2018

English: Bee macro
Date 18 July 2015, 10:35:52
Source https://www.flickr.com/photos/55293400@N07/20118812536/
Author Ömer Ünlü
Wikipedia.com - Creative Commons Attribution 2.0 Generic

Description: Honey bee
Date: 4 November 2011, 12:55
Source: Honey bee
Author AJC1 from UK
Wikipedia.com - Creative Commons Attribution 2.0 Generic

FARM BEE-KEEPING

E. E. TYLER and L. HASEMAN

The growing demand on the part of farmers and others for practical instruction in bee-keeping has called forth this report. It is not intended for the experienced bee-keeper, tho it may contain something new to him. It is intended primarily for the thousands of Missouri boys and girls and their fathers and mothers who want to know how they can get a few bees and properly care for them. The assistance of a practical bee-keeper, the senior author, has been secured to help prepare this report which has been made so simple that any child will be able to learn from it how to manage a few stands. Every effort has been made to discourage the waste of money in unnecessary fixtures and to place bee-keeping within reach of every Missouri family.

VALUE OF BEES

To man, the honey bee is the most valuable of all insects. Industry, its ruling passion, and instinct, which approaches and at times even surpasses human knowledge, have caused it to stand in the front rank of insects. If the bee did nothing more than furnish us with its matchless example of industry, social instincts and economy, its brief life would be well spent.

Every farmer should have two or more prosperous colonies of Italian bees. The honey-bee is our expert carrier of the pollen from flower to flower of fruits, vegetables and other crops. The fertilizing of one flower by pollen from another is the rule among honey plants and self-fertilization is the exception. If your fruit blossoms fail to set fruit perhaps a few colonies of bees would help. The two to ten dollars worth of honey from a colony of bees is a mere pittance of their real value on the farm. And yet from the point of view of honey production alone there is no legitimate enterprise from which a larger percentage of profit can be secured on the necessary capital invested and labor expended. No other enterprise fits in so well with general farming and offers such attractive inducements to the faithful and industrious.

THE LIFE AND HABITS OF THE BEE COLONY

The honey bee is closely related to the various other bees, such as the bumble bees, and to the ants and the wasps. Most insects of this type are armed with a poisonous sting, which is simply a modified ovipositor or structure used by the female for drilling holes in which to lay eggs. The male or drone bee is without the sting while the queen or fully developed female, and workers, which are undeveloped females, all have stings. This is also true of the other stinging insects.

Kinds of Bees in a Hive. There are three kinds of bees—queen, drone and worker. The queen is the female and the mother of the colony. It is her duty to lay the eggs in the cells provided by the workers. In a single day, when at her best, she may lay as many as three thousand eggs or twice her own weight of eggs. The workers feed her largely predigested food so that she can devote her entire energies to the production of eggs. Of all of the bees she is the most delicate. A sudden chill, jar or fright may cause her to stop laying. She is a wonderfully constructed machine into which food is poured and which grinds out great quantities of eggs.

FIG. 1.—DEVELOPMENT OF THE HONEY-BEE
On the reader's left are the earlier stages. a. Egg. b. Young larva. c. Old larva. d. Pupa. On his right are the three adult or mature forms slightly enlarged. a. Worker. b. Queen. c. Drone. (After Phillips.)

The Queen. The queen is reared in a special cell which usually points in a downward direction. It is about an inch long and as large around as the tip of the small finger. When swarming time approaches, the workers make one or more queen cells, each surrounding an egg previously laid in a worker cell, or the workers make queen cells, in each of which the queen may lay an egg. When the egg hatches the workers give the maggot-like grub a special food—royal jelly—which makes it grow rapidly into a queen instead of a worker. In about five days the grub is fullgrown and spins its cocoon, and the workers begin to seal up the cell. In this closed cell, within the cocoon, the grub changes to the pupa or resting stage, and on about the sixteenth day the adult winged queen comes forth.

In the meantime, under normal conditions, the parent queen has left the hive with a swarm. At first, the young queen is not much larger than the workers, but after she mates with the drone she assumes her full size, which varies from nearly an inch to a little more than an inch in length. In a few days after maturing, she usually leaves the hive on her mating flight, at which time she mates with a drone and then returns to become the queen and mother of the hive.

She mates but once, receiving from a drone a large supply of sperms which are stored in a special sac for use as needed in fertilizing her eggs during her life of usefulness. When she lays an egg, which is to produce a drone, the sperm is withheld, but it is permitted to enter eggs destined to produce workers or queens. This is a wonderful power possessed by bees, whereby an unfertilized egg produces the drone and a fertilized egg may produce either a queen or a worker depending upon the amount and kind of food received by the maggot-like larva which hatches from the egg. Any worker egg or larva under three days of age can be developed into a queen.

The Drone. The drone or male is produced previous to swarming or when the swarming fever occurs. The workers make special drone cells, which are about a fourth of an inch across and a half inch deep; and in these cells the queen lays a special so-called unfertilized egg which always produces a drone. The drone larva is full-grown and the cell capped about six and a half days after the egg hatches, and about twenty-four days after the egg is laid the drone appears. He develops less rapidly than the queen. The drone is both longer and broader than the worker and more than twice as heavy. Two thousand drones weigh a pound, while it takes five thousand workers to weigh a pound. The drone is about three-fourths of an inch long, clumsy in action and flies with a peculiar droning buz. He is not quite as long as the queen but broader and does not taper.

After the swarming season is over the workers usually kill off the drones to prevent them from continuing a burden to the colony; this is one illustration of their economy.

The Worker. The worker is the real bee of the colony. It plans everything and is the real master of the hive. It shows a kind of reverence for the mother queen but this is a case where the queen is ruled by her servants. During the late winter a strong colony may have only about fifteen thousand workers, but in midsummer there may be from forty to sixty thousand workers.

Each worker bee is a queen that was kept sexually undeveloped by the kind and amount of food given it while in the larval stage. The food given to the queen is of such nature that it develops the reproductive organs, while in the worker these parts are sacrificed for the greater development of the brain and the various organs used in collecting and storing food. At times so-called "fertile workers" appear, especially in weak or hopelessly queenless colonies. They may lay two or more eggs in a cell which produce drones, and as long as they are present the workers will not accept a true queen. To get rid of them unite the colony with a strong one as suggested on a later page in connection with uniting colonies.

The physical work of the hive is done by the workers. The young bee on first maturing serves for a time as a nurse. It eats honey and pollen and digests it, in part at least, and then regurgitates it for feeding the queen and the young brood. Later it may serve as a wax producer, secreting small flakes of beeswax, which are later worked over into comb. Finally it takes on the duties of a food gatherer. After circling about for a time to mark the place, it leaves for its first load of honey or pollen.

Pollen is gathered very largely in the early morning hours. Later in the day the pollen is dry and more difficult to get, while the nectar is in better condition for collecting. A study of the modifications of the parts of the legs for gathering pollen and the mouth parts for gathering nectar, will show that wonderful adaptations have taken place in this small creature.

The worker bees gather nectar, pollen, water and bee-glue or propholis while they secrete, by means of wax-glands on the underside of the abdomen, the wax used in the hive. The nectar is collected by thrusting the slender tongue down into the flower and drawing the nectar into its honey stomach where it is carried until the bee returns to the hive to regurgitate and store it in the cells of the comb. The bee deposits an acid secretion with the nectar and this secretion in time changes it into the real honey when it is sealed or capped. A worker bee will visit about twenty flower cups for one load of honey, which weighs about half a grain. About sixteen thousand loads are required for one pound of honey. A colony of bees will visit about three million flower cups a day.

The pollen is collected and molded into a small lump as it is being stuck on the outside of the hind legs, in the so-called pollen basket where it can be seen when the bee arrives at the hive. This is mixed with honey and other substances and is then known as bee-bread. Bee-glue is collected from freshly opened buds of trees and plants, and is also carried in the pollen basket. It is used to paint and stick fast frames, lids, and all parts of the hive, also to cover objects which cannot be removed from the hive, such as a dead mouse or snake. The wax is secreted in small flakes, by wax glands on the under side of the body, and is mixed with secretions so that it can be molded into combs.

The worker bee has no fear and yet will not "pick a quarrel." It is too busy to waste time in that way until it is forced to do so. Its aim in life is to feed, protect, and, if necessary, die for the good of the entire colony. Self-gain is unknown among bees and the whole colony works as a unit without friction or discord. Such harmony has never yet been attained by the human race.

The average life of the worker is about six weeks. It works from daylight until dark gathering stores, and then all night carries on the constant work of fanning the freshly stored nectar or thin honey to evaporate the excess of water. It literally works itself to death. When it is no longer able to rise at break of day and go in search of food, or when its wings have become so worn that it cannot carry a full load, it does not think of applying for a pension or a comfortable corner in the hive in which to end its days. Its work has been done and it throws itself from the hive, or drops in the field to die, in order not to pollute the hive or trouble the other workers. Those that drop dead while at work in the hive are quickly thrown out and replaced by others.

Races of Bees. The honey bee is not a native of America, but was brought over by the early colonists. There are a number of distinct races of bees. These vary in size and disposition, and are found in different parts of the world. In this country the golden or Italian bee is the favorite, but the black or German bee is also common. The black bee is usually found in the woods as it swarms oftener than the Italian. The beginner had better select the larger, more docile, golden Italian bee.

HOW TO GET A START

TO BUY ECONOMICALLY:

(1) Let a neighbor beekeeper hive a swarm for you. Requeen later if necessary to breed up.   Page 26
(2) Buy an old box hive from which you may transfer to a movable comb hive and requeen.   Page 19
(3) Buy a queen and few frames from some reliable bee supply man.   Page 40
(4) Buy bees by the pound with a queen.   Page 40
(5) Buy a complete hive of the best Italian bees delivered in place.   Page 23

TO GET BEES MORE ECONOMICALLY:

(1) Hive them from bee trees.   Page 19
(2) Catch your neighbors' escaped swarms by throwing dirt or water among them.   Page 25
(3) Anchor empty hives in forks of trees.   Page 23

NECESSARY EQUIPMENT

There is no need for the beginner to invest a great deal of money in equipment. One must buy a few things, but most of the necessary tools, or substitutes for them, will be found on every farm. The hives, frames, and sections should be bought in the flat and nailed at home. Wax foundation, two Van Deusen wax-tube fasteners, and, for an apiary of from five to ten colonies, a rapid reversible honey extractor and a solar wax extractor will be needed. Such necessary tools as hand saw, ladder, chisel, and pocket knife are usually at hand. A common butcher knife may be used for a honey knife or even for an uncapping knife. A few turkey or goose wings or quill feathers will suffice as a bee brush. Quart fruit jars are best for storing extracted honey. A five pound lard pail makes a convenient pail for wax-tube

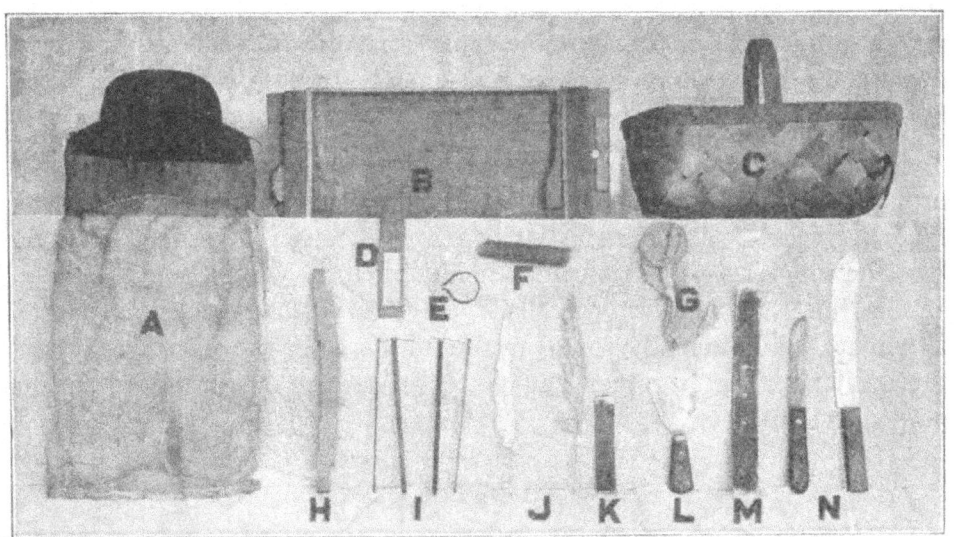

FIG. 2.—SOME HOME MADE EQUIPMENT

A. Bee-hat with band of black screen wire. B. Frame-nailing guide. C. Basket in which tools are carried. D. Queen-catcher. E. Pant guards. F. Queen-cage. G. Honey-straining bag. H. Frame for uncapping jar. I. Transferring sticks. J. Bee-brushes. K. Chisel. L. Glazier's tool. M. Hive tool made from buggy spring. N. Two butcher knives.

fasteners. A common glazier's tool is an excellent hive cleaning tool. Other necessaries, such as smoker, hat, swarm catcher, frame nailing guide, uncapping jar, queen catcher and cage, and hive tools can be easily made at home.

The Bee Hive. The movable comb hive is the only hive worthy of the name "bee hive." There are a number of kinds of such hives on the market. For Missouri conditions the Hoffmann-Langstroth or the so-called "Dovetailed Hive" has no superior and few equals. It consists of a bottom board, a body containing eight or preferably ten movable frames, a queen excluder, a super—the frames of which may be filled with sections—a layer of roofing-felt, and a hive cover.

Such hives can usually be bought in the flat in lots of three, five, or ten at a very reasonable figure. All parts and frames are standard and interchangeable. Nail and paint hives at odd times in the winter in order to have all ready for spring use. Then the larger part of the work of bee-keeping is done.

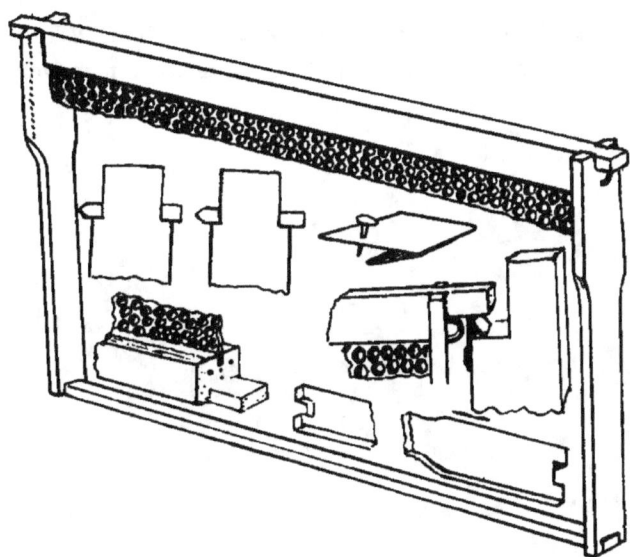

FIG. 3.—HOFFMAN-LANGSTROTH SELF-SPACING FRAMES
This illustration shows the foundation starter in place and details of construction.

Bee Smoker. Smoke is the bee-keeper's whip and should be used only in sufficient quantity to accomplish the end sought. The good points of a smoker are plenty of cool smoke when and where needed, and with as little expense, attention, and effort as possible on the part of the operator. On every farm there will be found discarded pails, kettles, and other raw material, from which one or more smokers may be made that will answer every purpose. In the illustration notice the draft inlet, also the extra twisted wire handles which keep cool enough to be handled with ease.

After the fire is started, the best fuel is a few corn cobs, and plenty of fine chips from the wood pile. Smother the blaze with green grass or leaves. Keep a smoldering fire, and the smoke will be of the best cold-blast variety, easily regulated by moving to or from the windward of the hive.

FIG. 4.—HOFFMAN-LANGSTROTH HIVE

The dove-tailed hive may contain either eight or ten movable frames.

FIG. 5.—FRONT VIEW OF HIVE

Note the bricks for raising the hive four inches from the ground. On the bricks is the bottom board, then the brood chamber, and between this and the super, the queen excluder. In the super is stored the surplus honey. Beneath the hive cover is a strip of tar paper or roofing-felt and on the top of the cover is a brick for weighting it down.

Bee Hat. The bee hat may be any stiff hat, preferably a discarded stiff-brimmed straw hat, with a band of black screen wire about five inches wide for a grown person. The band should be narrower for children, as it will ride on the shoulders if too wide. Sew this screen fast to the outer edge of the hat, and then sew a two foot band of cheese cloth or mosquito netting to the bottom edge of the wire band to tuck under the coat. Do not omit the black screen wire: it does not much obstruct the vision. It is a good plan to have two or three hats for the use of interested visitors.

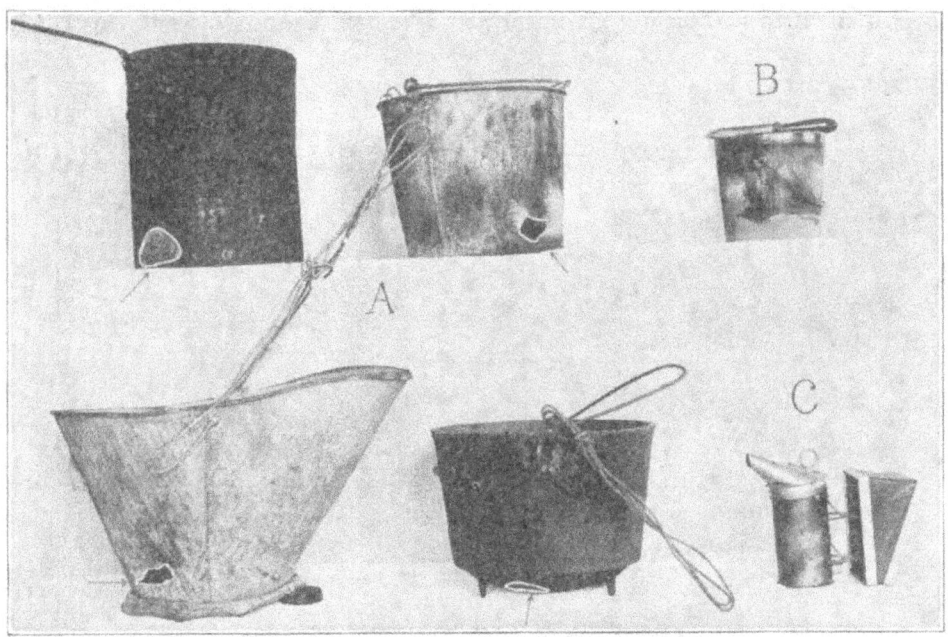

FIG. 6.—BEE-SMOKERS AND WAX PAIL
A. Four old pails with air vent near bottom. B. Wax pail with Van Deusen wax-tube fastener. C. Bellows-smoker.

Hive Tools. Very good hive-tools may be made of pieces of old vehicle springs ground to an edge on one or both ends, as shown in illustration.

The glazier's tool is the favorite hive tool. It will last a lifetime. To avoid damaging your frames always scrape or pry carefully when using strong tools.

Feeders. Following an unfavorable season or during a prolonged winter, a colony may use up all of its stores and need to be fed. Thirty pounds, or four full frames of honey, is all an average colony of bees will use during the winter. There are a great many types of feeders, but of all the feeders on the market the division board feeder is the most popular. It is used like a brood frame, and can be put into the brood nest in the same way. All that is necessary in feeding is to raise the cover and papers or cloth just enough to enable you to pour in the warm syrup from an ordinary coffee-pot or tea-pot. After doing this close up the hive, and the bees are thus supplied with food without exposing the cluster, and without the use of smoke. This feeder is especially adapted for early or cool weather feeding, or for weak nuclei at any time. It can be used when the bees could not take the syrup from other kinds of feeders. It is made in two sizes, one the width of a brood-frame holding a pint, and the other two inches wide holding five pints. The small one is used especially

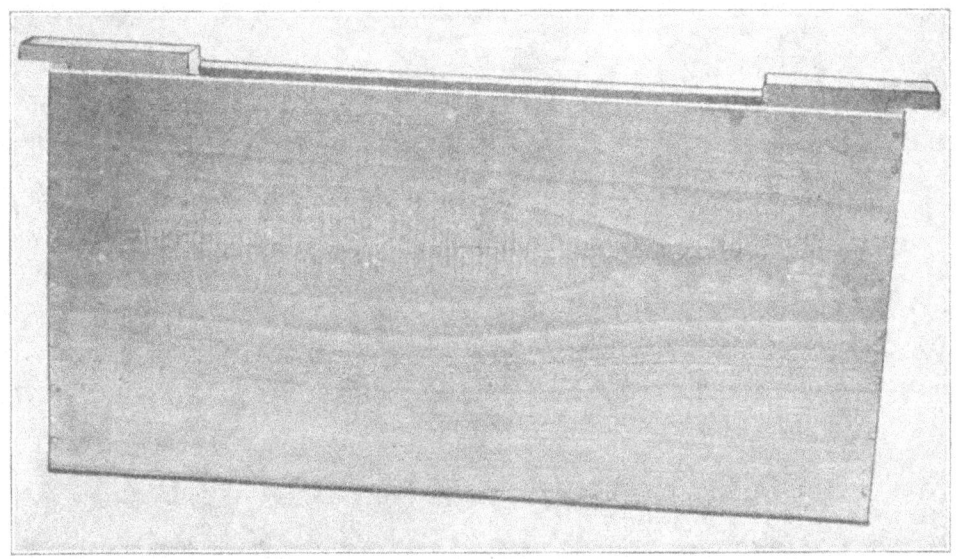

FIG. 7.—DOOLITTLE DIVISION BOARD FEEDER
Especially adapted for early or cool weather feeding.

for weak nuclei in queen rearing. One quart of boiling water into which two quarts of good granulated sugar is stirred, and all brought almost to the boiling point and given to the bees while warm is nearly as good as honey for feeding bees.

Uncapping Can or Jar. Beeswax is valuable and all scrapings from hives, and cappings cut from combs at extracting-time should be collected in a can or jar. An uncapping jar can easily and cheaply be made by using a five gallon stone jar or one of larger size if preferred. Take a piece of inch board long enough to reach entirely across the top of the jar. At both ends notch down about an inch, so the board will set in the jar, fitting closely to hold it firmly in place. Notch the top as shown in figure and put four sharpened nails in it at a distance to suit your frames. These prevent the frame of capped honey from slipping when it is standing on end for uncapping.

Honey Extractor. After one has six or more productive colonies of bees he should secure an extractor, especially if he is working to get extracted honey, which is usually advisable. An extractor is provided with rotation baskets in which the uncapped combs of honey are placed, and as the combs are rotated the honey is thrown out. For this reason extracted honey is sometimes incorrectly called "slung honey."

There are a number of different types of extractors. A small hand extractor of the rapid reversible type is sufficient for an apiary of from twenty-five to fifty colonies. This will cost about $15.

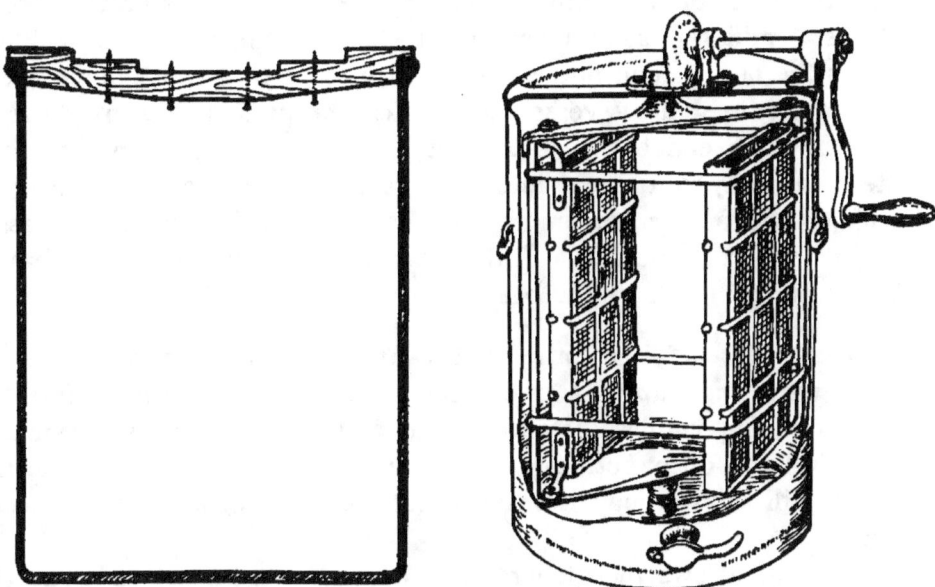

FIG. 8.—UNCAPPING JAR AND RAPID REVERSIBLE HONEY EXTRACTOR

A five-gallon jar furnishes a convenient and cheap apparatus for uncapping. Uncapped combs of honey are placed in the baskets and rotated.

Wax Extractor. By saving all cappings and trimmings from bur and spur combs that would otherwise be lost, you will soon be able to buy a solar wax extractor. An extractor may be made from an old stove pan or dripping pan. Cut a hole in one end of the bottom and put the chunks of compressed comb in the other end. Place it in the stove oven with the hole out over the edge of the oven and set a pan, containing a little water, for the wax to drip into as it melts. Leave the oven open and put a brick under the back edge of the pan. Use a moderate fire and as the wax melts and flows out, fresh comb may be added. When cool the wax will be ready for market. A sheet of tin twenty inches wide and twenty-four inches long with the edges turned up will make a good oven extractor. The deposit left in the pan extractor is called "slum-gum." It has a low market value but is of use as fuel. If the wax is saved it should bring enough money to buy all necessary hives and supplies—even a new queen now and then.

The old daub method of extracting beeswax is still used by some bee-keepers. Old combs are crammed into a cloth flour sack and boiled in a kettle of water until all the wax is dissolved. When cool the wax can be removed.

Beeswax rendered in a solar wax-extractor has the finest color and will sell better than wax rendered in other ways. A convenient solar wax extractor can easily be made by taking a shallow cracker box, or larger similiar box, and bending a sheet of tin so that it will fit down in the box. Bend one end of the tin to form a spout and under this place some shallow vessel into which the melted wax may drip. Raise the back end of the box enough to face the sun so that the melted wax may run down into the vessel. Place the old combs in the upper end of the box on the tin, cover all with a pane of glass, and the heat from the sun will do the rest. By this means you have the wax out of reach of the bees.

Swarm Catchers. A convenient swarm catcher is needed when a swarm settles high up in a tree. One of the best is a super, full of empty extracting combs, fastened on the end of a pole or two-by-four. By placing the open bottom of the super close against the swarm, the bees will soon occupy it when it can be taken down and the bees hived.

Other good swarm catchers consist of (1) a bushel feed-basket with two frames, containing comb, securely wired in the basket, fastened up side down over the end of a pole; (2) a nail keg inverted over the end of a pole and nailed to it with a comb tied inside as a decoy; (3) two market baskets tied facing each other with some comb inside; (4) a light goods box with a comb fastened inside. In case the surrounding trees are tall a ladder also may be needed but if the bees are located in a grove of tall timber it is better to raise queens and draw swarms as directed later under artificial increase.

Queen Catcher. The queen bee is a most delicate creature and must be handled with care. An experienced beekeeper can pick up and handle a queen with safety but the beginner had better use a queen catcher. A convenient homemade catcher can easily be made by taking a piece of soft pine seven inches long, one and a half or two inches wide, and at least a quarter of an inch thick. By means of a pocket knife cut out a hole in it near one end, three and one-half inches long and an inch and a quarter wide. Then take a few small tacks and fasten a strip of screen wire over one side of the hole. Bevel the edges of the piece of wood and take a piece of tin and turn down its edges so that it will slip on at one end and slide along so as to close the lower side of the hole cut in the wood. When ready to catch a queen slip the tin a little and place the hole over the queen and then gently slide the tin back, enclosing her in the trap.

Introducing Cage. Bees will not always accept a new queen. It is best to introduce her in a cage that she may be protected while they are

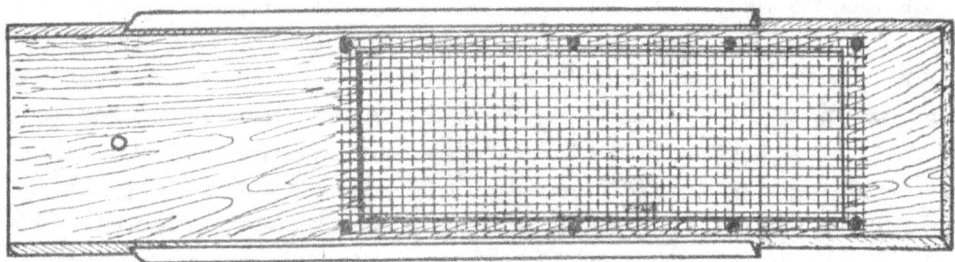

FIG. 9.—QUEEN-CATCHER

Above a strip of screen wire and below a sliding strip of tin is used for closing the opening.

FIG. 10.—SMALL BENTON QUEEN-SHIPPING CAGE

Such a cage is necessary when queens are shipped and it may serve as an introducing cage.

getting acquainted with her. A convenient introducing cage is made by cutting a piece of screen wire about four inches wide and five inches long and rolling it around your finger. Tie some frayed strands of wire around it, crush some empty comb in one end, and after the queen is in close the other end in the same way. The shipping cage in which a queen is received by mail will also serve as an introducing cage.

Transferring Sticks, Splints or Wires. It is absolutely necessary that you have something to hold the pieces of comb in the frames when you transfer bees from a tree or gum to a hive. Cut out seventy-two sticks of straight white pine, or any other good wood, making them about an eighth of an inch thick, three-eighths of an inch broad, and nine and a half inches long. About an eighth of an inch from each end cut a small notch on each side deep enough to hold a wire safely. Cut seventy-two pieces of copper wire each three and a quarter inches long. Wrap one end of each wire securely to each end of thirty-six sticks, so you have two and a quarter inches of loose wire to extend across the frame to tie to the splint on the opposite side. When done you should have thirty-six sticks with a wire on either end, and thirty-six without wires. This is enough for any ordinary transfer and they will last a lifetime.

Pieces of number fourteen wire, bent in the shape of an L at each end and long enough to span the top and bottom bars may be used for the same purpose. Brads or wire tacks may be used quite successfully to hold the splints in place. Common twine or binder twine is often used but it is easy to make a botch of it. Do not try it but make the wood splints.

Frame Nailing Guide. The hives and frames should be bought in the flat and nailed together at spare times during the winter. A convenient frame nailing-guide can be simply made as shown in the illustration and when such a guide is used all frames will be exactly alike and they can be much more firmly put together. A box, a few nails, a few screws, a short spiral spring and a pocket knife, screw driver and hatchet are all that are needed.

HANDLING BEES

The most important and the most interesting of all beekeeping is the actual work of handling them. One must be able to open a hive and follow their work. The timid and inexperienced may have a little trouble at first, but with a little patience, care, and composure one soon becomes fascinated with this part of beekeeping and then there is danger that one will be opening the hive too often. One must be careful about this and learn to follow the work of the colony from outward signs.

External Examination. It is not well to disturb the bees too often. Let them alone as long as they collect pollen in the morning and a large number of bees pass in and out all day coming in laden with nectar. To evaporate this the bees pass currents of air over the combs causing a "roar" that can be heard two or more rods away. As long as these prosperous conditions continue it is best to look after the supers, keeping the surplus honey removed and plenty of empty combs above, or the bees will be forced to store honey below and crowd the queen for room by too much prosperity in the wrong place. Use two supers if the flow is plentiful with the empty one next to the queen excluder.

FIG. 11.—FRAME-NAILING GUIDE

The left-hand end bar of frame to be nailed has the sharp V-edge up while the right-hand end has it down. The bottom bar is a quarter of an inch shorter than the top bar. This brings the bottom of the end bars a little nearer together which helps to prevent the crushing of bees when the frames are raised or lowered in the hive.

If the bees suddenly stop carrying pollen or their enthusiasm is dampened, they are probably queenless. Requeening is more fully discussed later. On the other hand, if they "hang out," try more shade and remove surplus honey. If this does not start them to work when other colonies are storing honey take out two frames containing the most honey from the brood chamber, and replace them with empty frames containing starters or full foundation. Remove all drone comb, also reverse the ends of every other frame, trimming off thick places if necessary. If you notice any comb that is too one-sided, cut along the frame and press it in line. Give the two full frames to some colony that needs building up and both hives will soon be doing a thriving business; the first one pulling comb which the queen will lay full of eggs, forcing surplus stores above; the second being strengthened so that the bees usually enter the super without further trouble.

How to Handle Bees. In handling bees courage and confidence with steady accurate movements are very essential. "The brave die but once, the coward many times." To begin, fire your smoker, put on your bee hat, tuck the cloth or netting under your coat, and put on a pair of bicycle pant guards, or a pair of shoe strings will answer the purpose. Take your smoker in one hand and a market basket containing bee brush (quills), butcher knife, hive and cleaning tools and pocket knife in the other. Proceed to the windward of the hive and disperse the guards at the entrance with a little smoke, remove the cover and cloth or paper, after letting a little smoke pass underneath to turn the

FIG. 12.—QUEEN-EXCLUDERS.
Bound perforated zinc excluders for eight- and ten-frame hives. The openings are small enough to prevent the queen getting thru to lay eggs in the supers, but the workers get thru readily.

bees to their stores. Then set smoker aside and scrape the top of the frames if they need it. If a super is on, first remove it by carefully prying and cutting it loose from the queen excluder with a strong butcher knife, so as not to injure the queen excluder. Then carefully remove queen excluder and begin at one side of the brood chamber to pry a frame over and up slowly at both ends. Remove it by taking hold of both ends. Lift it steadily and evenly and if too tight pry it a little higher and raise it up about half way. Hold it in place with one hand, and brush adhering bees downward gently into the hive as you gradually remove the frame. Place it on edge against the outside of the hive. Pry the next frame over toward the vacant space, lift it, and examine for honey and brood. Note condition carefully, set frame in place of first one, and continue until you have seen all of the combs, and possibly the queen. Note her size, color, and movements.

When you have more experience it will be no trouble to handle frames, two and three at a time and learn all that is necessary about the condition of a hive in ten minutes or less.

Finding Wild Bees. Every farm boy if he has keen eyes can easily find bee-trees. Before starting on such a hunt first make a syrup of sugar and boiling water about as thick as honey or simply use honey. A pint will be sufficient. Take a few matches, a shingle, some old honey comb, half a dozen pins, and two wide topped glasses or jelly jars and proceed to the windward of the woods to be tested.

Select an open place and make a trash fire, dropping a little comb on it to attract the bees. If it is clear and warm, in less than fifteen minutes bees will be seen to follow the smoke to the burned down fire. Pin two pieces of comb on the shingle so that they can be covered with the glasses and pour each piece full of syrup or honey. Watch the flight of bees as they leave until a definite line is established. Then cover some of the bees with the glass and carry them at right angles to the first line and liberate them. Soon a second line will be formed and marked as before. The place where these two lines cross will locate the bees. Bees always fly in a straight line when they leave and when they return home and it is easy to find them by following their lines of flight. The term "bee-line" comes from the fact that bees fly in a straight line. When you cut a tree, watch for new lines as there are probably more close by.

Some use a wooden box with a draw cover about the size of a cigar-box and pinned or tied pieces of comb in the bottom of the box. In this way one is not bothered with the jelly jars. The honey or syrup can be carried in a bottle in the pocket.

A little flour sprinkled on the bees enables one to see them better and at the same time saves syrup. Anise oil has scent attractive to bees and by putting a few drops on a cloth and hanging it up the burning of the comb will not be necessary. The best time for hunting wild bees is before any bloom appears in the spring. In the summer the best way to locate them is to find where they water and mark their lines of flight as described above. When a tree is located mark it, see the owner, and have a "bee" transferring as described in the section on transferring bees. Bee trees can be found only on warm days, when the bees are out.

Transferring Bees from Box Hive or Tree. One can often buy bees in old box hives or gums at a low price and afterwards transfer them to movable comb hives. Bees found in trees or elsewhere can be handled in the same way, so it is necessary that one be able to transfer combs and bees into the particular hive desired.

The best time to transfer bees is in the spring when fruit trees are in bloom. At that time there is less brood and honey to contend with and the weather is more settled. The best time of day is before six o'clock in the morning, tho any part of a warm day will do.

Smoke the bees and spread a cloth in front of box hive as for hiving a swarm. Tip the box hive forward on the cloth with the combs

FIG. 13.—PARTLY OPENED BEE-TREE
Most of the bees are clustered around the knot-hole entrance along the side of the log.

on edge. Set the new hive on the old stand with frames near by. Lay a board a little larger than a frame on a goodsbox for a table, spread a cloth over the board, and lay a frame on it with four wired splints underneath. Use the short buggy spring hive tool for cutting nails and prying, and the butcher knife for loosening the comb. Smoke and brush the bees back. Cut out the comb, fitting it closely in the frame until full and fasten on splints. Tip the board up behind the frame bringing all on edge. If all is straight, and well balanced, set it in the hive. Continue in like manner until done. Remove all drone comb and any excess of chunk honey for table use. Chunk honey is described later.

Within a week to ten days all splints should be removed and a queen excluder with a super placed above. Then cover with paper, or roofing-felt and weight the hive cover with a brick.

Transferring from a bee tree is much the same with the exception that the hive of bees must later be moved. If you cannot get near the place with a wagon, rope the hive to a pole and carry it out. Further suggestions may be found under "Moving Bees".

FIG. 14.—CARRYING BEES OUT
These were transferred from bee-trees and after a few days brought out of the woods.

When starters are not used in the frames the bees often build their combs in almost any shape. Then it is necessary to cut out the comb and fit it in frames, the same as when transferring from a tree or box hive.

Removing Bees From Objectionable Places. In nearly every locality a swarm of bees may at sometime get into the side of a granary, smokehouse, dwelling, church, or occasionally even into the trunk of a valuable shade tree, low enough to make themselves troublesome to both man and beast. The best time to remove them is in the spring or summer. Stop all holes, except the main entrance, with rags covered with a clay-mud coal-oil mixture. Make all but the main entrance absolutely bee-proof, thus compelling the bees to come and go by the main entrance.

Next take the end out of a cracker box and bore a hole two inches in diameter thru the center. Over this make a screen wire cone, as in a fly trap with the small end about the size of the small finger. Over this cone make a second larger cone, and preferably a third over the second, as the bees hunt carefully and might escape thru the trap back into the house if only one cone were used. Take a second block of wood, two inches wider than the cracker box end and six or more inches longer. Bore a hole thru the center as before and nail the cracker box end across it so that the holes match and the cones stand upward. Then nail the rest of the cracker box back on the end so that the wire cones projects into the box. Nail some pieces in the box for the bees to cluster on and put screen over the top for light and ventilation. The trap is now ready to fasten over the entrance. Put it up before daylight when most all the bees are inside, and be sure to stop up all cracks behind the block. Put screws thru the larger block in fastening the trap to the house, since you should not drive nails, and support the end of the trap by running bailed hay wires around it and fastening these to screw eyes or nails higher up on the side of the house. In the morning, when the bees come out, they will pass thru the hole in the blocks, thru the cones, and find themselves trapped in the box. Sprinkle them every hour or two with honey-water or sweetened water and make sure that they have shade. In the evening thoroughly drench them with the sweetened water, and remove the trap, and hive the captured bees on a few frames with combs containing eggs and sealed brood and close up the entrance of the hive as in moving bees. Replace the trap for the next day's catch and continue until all are trapped. As soon as they start queen cells in the new hive, they may be permitted to go back in and rob the old stand, carrying the honey to the new hive. The queen will then starve. Later close the hole and the bee moth will clean out the combs.

If the colony of bees is not wanted, simply use the blocks and cones for trapping the bees out, and as they cluster over the cones and blocks, drench with boiling water until they are all destroyed.

Where it is possible to quietly cut or draw the nails, some of the siding may be removed and the bees and combs transferred, as in case of a bee-tree or box-hive. If the bees are not too troublesome, this can be deferred until cold weather when the bees are less active. After removing them, stop all holes with small pieces of tin cut from tin cans.

Moving Bees. In moving bees, close the entrance with screen wire before day, when all bees will be inside. Take a strip of screen wire three inches wide and two inches longer than the bee entrance, bend over the ends so that it is exactly the length of the entrance, then bend the strip lengthwise into a V-shape, and push it tightly into the entrance. This closes the entrance and serves for ventilation. Be sure that there are no other openings left. Run a baled hay wire lengthwise around the hive drawing and twisting it up tightly. Run another crosswise in the same way. Have a wagon close at hand with plenty of hay, straw, or small brush to relieve the jar. Set your hives crossways, far enough apart to crowd a partly filled sack of straw, leaves or brush between them and the sides of the wagon bed. When all is carefully done, hitch your team to the wagon and drive to where the bees are to be placed. Unhitch before doing anything else. Then set all the hives in permanent place as nearly a rod apart as convenient. Take off the baled hay wires. Place a wisp of loose hay, straw, grass or fine brush close up in front of the entrance and open the entrance about two inches at first. The trash in front of the entrance causes every bee to take notice and mark the new location. If the entrance were thrown wide open, the bees would come out too fast and soon find themselves lost in mid-air; and if not too far removed from the original place, would go back and find themselves homeless. Move bees in the cool of the day. Either single queen or a carload of stands can be transported with ease and safety.

Uniting Weak or Queenless Colonies. Sometimes there are two or more weak colonies, which would perish during the winter if left separate, but if united would form a strong colony and winter in good shape. To unite them cut a strip of common fly screen the width of the hive and three inches longer than it. Tack strips of plaster lath securely along the two edges and one end of the screen. Then turn it over so that the strips are all above, and tack a thin three inch strip under the other end which will later serve for an alighting board.

If the two colonies to be united are far apart, the one had better be moved over beside the other and elevated at the same time, as in moving bees. If, however, they are near each other, each evening one may be moved a foot or two nearer the other and when about eight feet away the hive may be elevated and brush put below to make the bees note the change in elevation. Leave them side by side for three days, after which remove super and queen excluder from the stationary stand and put the screen over it so that it fits bee-proof with alighting board in front. Then lift the other brood chamber from its bottom board and set it on the top of the screen. In three days the queen may be removed from the upper colony, if a queen is present, and in three days more the bees will all have the same odor; when the screen may be removed and the queen excluder replaced. The colonies will then be as one, all going and coming by the main entrance below and working in unison. In the same way a greater number of weak colonies may be united in succession. In about three weeks all the bees will have hatched from the upper brood chamber, when it may be replaced with a regular super and its combs removed for rendering or for saving as desired. With this method the bees do not stop working and there need be no worry about robbing or fighting. As a protection, the entrance of weak colonies should always be contracted, the size of the entrance depending upon the strength of the colony.

If the entrances of the two hives of the colonies to be united face in opposite directions, the direction of the one can be changed by one-fourth turns every two or three days.

Reviving Weak Colonies in the Spring. Sometimes in the spring a colony may be slow in starting to build up, and yet have a valuable queen. With a little attention such a colony may be easily revived. If discovered early while the weather is cool, contract the entrance and place it on the top of some other strong colony with the uniting screen between and with the entrance the reverse of that of the other colony. Put in a division board feeder full of warm syrup as in feeding bees. The warmth of the lower colony and the food quickly revives the queen and the entire colony. This is the simplest and most practical method of reviving a weak colony. Afterwards it should be removed to a new place, as directed under "Moving Bees".

If this does not give the desired results, the queen is worthless and should be destroyed. The colonies should then be united and later in the spring divided if necessary.

INCREASING COLONIES

There are two methods of increasing colonies of bees: the natural method of permitting them to swarm naturally, and the artificial method of dividing or forming nuclei.

Natural Swarming. The honey bee is a social creature and always lives in colonies. The number of bees in a colony increases until a limit is reached. When the hive, tree, or other home of the colony becomes too crowded, it prepares to divide, making two colonies.

There are other conditions besides overcrowding which bring on a division of the colony, or swarming. Swarming is the natural method of increase of bee colonies and where it is entirely under the beekeeper's control it is considered the best method.

On preparing to swarm, one or more young queens are developed by special feeding and an abundance of drone brood is produced. Then a few days before the young queen or queens emerge, the mother queen is literally carried from the hive by the workers and forced to go with the swarm. As a rule, about half of the bees go with the swarm and they usually cluster near the hive before starting on what may be a long flight to some tree or empty hive. If not hived at once after clustering, the swarm may leave for its new home which has been previously selected and cleaned out.

Hiving a Swarm of Bees. Have hives ready and work quietly when a swarm issues. Wet the inside of the hive with strong brine, or cool water. Put it carefully in place on bricks with the front slightly lower. Spread a cloth in front so as to guide the bees into the hive. By this time the swarm ought to be settled on a branch, which can be cut off, taken to the hive, and the bees shaken on the cloth. To start them in push or brush a few up to the entrance. Once started watch for the queen to make sure of her safety. Sprinkle the swarms with water at once after settling as this will decrease the possibility of escape and make them easier to manage.

Sometimes a swarm will settle high up on the trunk of a tree or other place difficult to reach, when the swarm catcher will be needed. To make sure that they will stay give them a frame or two of brood, and place the hive so that it gets only the morning sun and the colony will go to work at once.

Artificial Increase. Most beekeepers prefer to increase their colonies artificially rather than run the risk of losing a few good colonies by permitting them to swarm. Some attempt to prevent the escape of swarms by clipping the queen's wings. If one wishes to increase by forming nuclei he can always rear queens from his best stand for the nuclei and thereby build up his apiary. The majority of the experienced beekeepers prefer this method, but many beginners prefer to let their bees swarm in the natural way.

How to Form Nuclei. By a nucleus—plural nuclei—is meant a start from which a full colony may develop. In warm settled weather when honey is plentiful, a single brood frame containing fresh eggs, well covered with bees, set over to one side in an empty hive with a frame rich in honey next to it, and closed up tightly with grass for four or five days, will build a few queen cells, become contented and raise a queen. For protection the entrance should be closed up so that only two or three bees can pass out or in at a time. It may later be enlarged as needed. In farm practice strong nuclei or new colonies may be formed by simply dividing prosperous colonies when honey is plentiful. Set over in a new hive, three or four frames with adhering bees and queen, and fill the rest of the two hives with frames containing foundations or starters. Of its own accord the old colony will rear a young queen and under favorable conditions soon build up a strong colony. The entrance of the new hive should be stopped up tightly with grass or most of the bees may return to the old hive. If in four days they have not gnawed out, help them, and as the colony increases enlarge the entrance. This, too, will develop into a strong colony.

REQUEENING

From time to time old queens or inferior queens must be replaced, therefore the beekeeper should know how to secure strong queens when needed. Queens can be bought from those making a specialty of queen rearing, but there is always more or less uncertainty of getting them when needed. It is really better to produce your own queens, if you have one or more choice colonies from which to rear.

If you wish merely to replace an old worthless queen or requeen a queenless colony, you may simply kill the old queen and after five or six days go in and cut out all queen cells which the workers may have started. Take from your best colony one or two frames rich in young brood and eggs and put them into the queenless colony in place of two of their central frames. In time they will develop a queen, the hive will take on a new life, and the colony will develop into a prosperous one.

Introducing Queens. Be sure that your colony is queenless. If you kill the queen yourself, you may introduce immediately, tho it is safer to wait from twenty-four to forty-eight hours. A new queen should never be turned loose at once in a colony, as she does not have the same odor, and the workers are almost sure to "ball" and kill her. The cage in which she is received will serve as an introducing cage, or she may be carefully transferred to the introducing cage already described. In place of a mail-order queen, she may be a home product taken from one of the most prosperous colonies with a view of having that colony produce a young, thrifty queen. In that case the queen is caught in the queen catcher in which she may be introduced, or she may be transferred to the rolled screen wire introducing cage, the ends of which are closed with plugs of comb.

To introduce the queen open the queenless hive using smoke and giving the bees time to fill. When thoroly subdued space the top bars near the middle of the hive to suit the cage. Then place the cage between the combs about six inches from the back, directly over the main cluster and in the most favored place in the hive. Close the hive and success is almost sure, but it is best to peep in after about two or three days. If she is not yet out, break up the plug of comb with the small blade of a pocket knife, loosening it so that in a few hours she may escape.

By the next day she will probably be the queen of the hive ready to begin laying at once. Remove the cage for future use.

Never breathe on bees or queen, and never expose the queen to cold or draft.

After you have once succeeded it is easily done. If you should not succeed the first time, try again. A bright boy or girl of twelve can soon learn to introduce queens in this way.

HONEY PRODUCTION

Honey is produced in three forms—extracted, chunk, and comb or section honey. A colony will produce much more extracted honey than either comb or chunk honey. The secretion of wax, used in making comb, is slow and expensive work. It takes from fifteen to twenty pounds of honey—used by the bees as food—to produce one pound of the wax; so for greatest honey yield the combs should be saved and returned uninjured for refilling. The average yield of chunk honey or comb honey is about twenty-five pounds per colony; of extracted honey from fifty to seventy-five pounds. The beginner with a few colonies may not feel able to buy an extractor, but often he can co-operate with neighbor beekeepers and secure an extractor for the neighborhood.

Extracting and Extracted Honey. Extract only ripe honey which will always be found sealed or capped. A little honey not yet fully ripened and capped mixed with other extracted honey will spoil the whole lot. Before extracting the honey must be uncapped. To uncap honey the uncapping knife with a blade eight inches long is best, tho a good butcher knife will do. Keep it as sharp as a razor. Stand a frame full of capped honey on end on the board of the uncapping jar. The sharp nails will keep it from slipping. Shave caps off in large thin sheets letting them fall into the jar. The low places can be readily uncapped with the rounded end of the blade. Turn the frame and do likewise to the other side. It is now ready for the comb pocket of the honey extractor.

Before using, clean the extractor thoroly and oil the bearings sparingly. Place it on an inverted box high enough for a pail or jar to be placed under the honey gate and anchor it securely. Close the honey gate. Place an uncapped frame in each of the comb baskets and partly extract from one side. Reverse the baskets and extract part from the other side so as to avoid cracking the combs and then speed up and clean the one side and reverse the baskets and clean the other side of both combs. A little practice will soon indicate the speed required for each part of the work so as not to injure the combs. Extracted honey can be stored the same as canned fruit.

By taking the capped frames from one super at a time, shaking or brushing the bees back, the honey will be warm enough to flow well and the comb in about its toughest state for extracting. Return

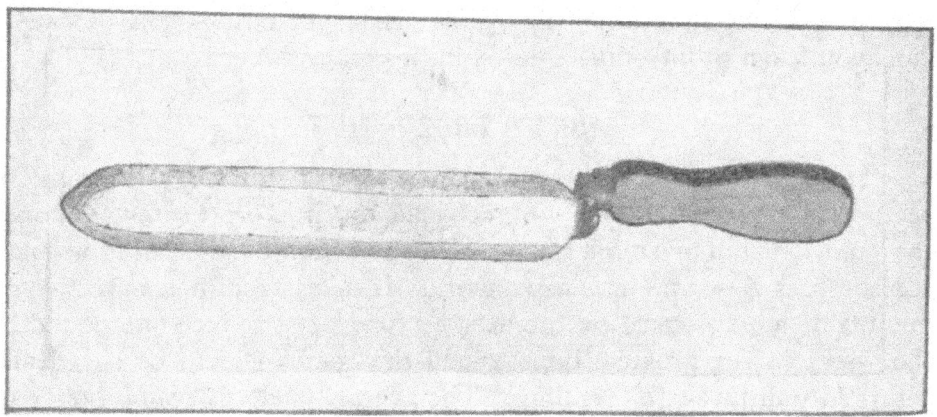

FIG. 15.—UNCAPPING KNIFE
Low or depressed places can be uncapped with the rounded end of the blade.

the empty combs to the super and in this way go thru all the supers. If the flow is plentiful it may take two or even more supers to each hive to keep up with the needs of the bees.

Extracted honey should always be strained thru a cloth to remove pieces of wax, chips or other particles. To strain honey easily, use a piece of cheesecloth tied or pinned over the vessel. Another convenient way is to make a bag about the size of a five cent salt sack with a twisted wire ring sewed around the top to hang it under the honey gate. A wire ring one foot across covered with cheesecloth so it will bag some is handy and easily cleaned. By means of a stick under one edge it can be held up, while the other side rests on the edge of the vessel.

Chunk Honey. Chunk honey is obtained by simply going into the hive and cutting out chunks of honey and comb, using it in that form.

Comb or beeswax is very indigestible and should not be eaten. Chunk honey should be put in a pail and this set in a vessel of scalding water until the comb melts. Then on cooling slightly the wax will form a cake over the honey, which may be taken off leaving the pure honey ready for use after straining.

Comb Honey. Fancy market honey is usually produced in small sections each weighing about a pound and when properly cared for it can be produced profitably. Some beekeepers work for section honey

FIG. 16.—SHALLOW EXTRACTING SUPER
It may take two or even more supers per hive to supply the needs of the bees.

only but it is usually more difficult to get bees into a super of sections than into an extracting super. To overcome this, first get them to partly fill a super of frames for extracting and then raise it and put a super of sections with foundation starters between it and the brood chambers. They will then usually store in the sections.

HONEY AND BY-PRODUCTS

Pure honey is a most wholesome and delicious sweet and its more general use should be encouraged by a more generous supply on the market as well as on the table of the beekeeper. The real value of honey for table use is well known and need not be emphasized. It is much more delicious and wholesome than any syrup or similar material used on the table. It can be used in place of sugar, sorghum, and other syrups in many recipes and when used it improves the flavor of the finished product. More honey should be produced and used on every farm. As there are very few recipes calling for honey it is thought best to give a few tried ones.

RECIPES FOR THE USE OF HONEY

Pork Cake. Two cups of sugar, one cup of extracted honey (or sorghum), one cup of sour milk, one pound of pork minced fine, one pound of raisins, four eggs, one grated nutmeg, one teaspoon of soda, one tablespoon of cinnamon. Stir same as fruit cake. May be kept six months.

Soft Gingerbread. Half a cup of sugar, one cup of extracted honey (or sorghum), half a cup of butter, one teaspoon each of cloves, ginger, and cinnamon, two teaspoons of soda dissolved in one cup of boiling water, two and one-half cups of flour. Add two well-beaten eggs the last thing before baking.

Honey Muffins. One pint of flour, two teaspoons of baking powder, one-half teaspoon of salt, sifted four times; yolks of two eggs beaten lightly; one and one-fourth cups of cream. Beat thoroly, then fold in lightly the beaten whites of two eggs and two tablespoons of extracted honey. Bake in muffin pans and serve while hot.

Honey Cookies. One large cup of extracted honey, one-half cup of butter, two eggs beaten lightly, one teaspoon of soda, dissolved in a little warm water, flour enough to make a soft dough. Flavor with nutmeg or vanilla. Pinch off small pieces of dough, rolling them in balls, flatten slightly, but do not crowd in the pan. Bake in quick oven.

Christmas or Honey Cookies. One pound of honey, one pound of flour, one-half pound of butter, one-half pound of almonds chopped with skin on, grate rind of half a lemon, one teaspoon of cinnamon, a little ground cloves. Let the honey and butter come to a boil, add flour, spices and almonds, cool and roll.

Brown Bread. One-half cup of honey, one and one-half cup of sour milk or buttermilk, three cups of graham flour, one teaspoon of soda. Nuts and raisins may be added if desired.

Honey Apple-Butter. One gallon of good cooking apples, one quart of extracted honey, one quart of honey vinegar, one teaspoon of cinnamon. Cook apples smooth, add vinegar and honey when done, stir in cinnamon. It will be clear apple butter of fine quality easily kept in jars.

Chocolate Caramels. One pint of sugar and one pint of extracted honey (or sorghum), one-quarter pound of grated chocolate, one-half cupful of sweet cream. Try often while it is boiling, by dropping a little in cold water. When done pour about one-quarter inch thick on greased tins.

Butterscotch. One cup of butter, two cups of sugar, two cups of extracted honey (or sorghum), one heaping teaspoon of cinnamon. Boil ten minutes, pour in a buttered pan, and when cold cut in squares.

Walnut Creams. Boil to the hard snap stage one cup of grated chocolate, one cup of brown sugar, one cup of extracted honey (or sorghum), one-half cup of sweet cream. When it hardens on being dropped in water, stir in butter the size of an egg. Just before removing from fire add two cups of finely chopped walnuts. Stir thoroly and pour on buttered plates to cool, cutting it in squares. Other kinds of nuts may be substituted.

Cracker Jack. One cup of brown sugar, one cup of extracted honey (or sorghum). Boil until it hardens when dropped in cold water. Remove from stove and stir in one-half teaspoon of soda. Stir in all of the popcorn it will take, spread on greased tins and mark in squares.

Honey Vinegar. Dissolve thoroly in two gallons of warm, soft water one quart jar of extracted honey. Give it air and keep in a warm place where it will ferment and make good vinegar.

Boston Baked Beans. Soak one quart of beans over night. Drain, cover with fresh water and heat slowly keeping below the boiling point. Cook until the skin will burst when a few are removed in a spoon and blown upon. Then drain. Scald the rind of three-fourths pound of fat salt pork, scrape, remove one-fourth inch slice and put in bottom of bean pot or baking dish. Cut thru rind of remaining pork every half inch. Put beans in pot, bury pork in beans, leaving rind exposed. Mix one tablespoon of salt, one-half tablespoon of mustard and four tablespoons of honey; add one cup of boiling water and pour over the beans. Add enough more boiling water to cover the beans. Cover bean pot, and bake in a slow oven six or eight hours. Add more water as needed.

SOME USES OF BEESWAX

Next to honey, beeswax is the most important product of the bee. Its commercial value is considerable. About the apiary it is of use as foundation, for fastening in foundation, sealing cracks and leaks, and for many other purposes. Melted beeswax is molded into thin sheets and pressed between dies to form foundation. This is fastened into frames and is the midrib of comb for the bees to pull up the edges forming the completed comb. Artificial comb is not manufactured at present.

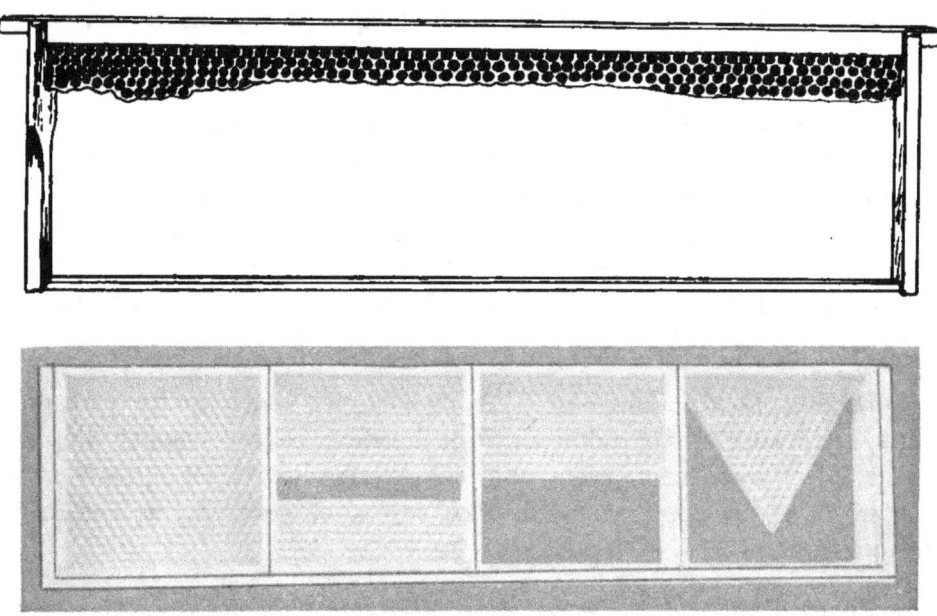

FIG. 17.—SHALLOW EXTRACTING FRAME AND SECTIONS
Comb foundation should always be used in frames and sections. In sections, strips of different shapes may be used.

There is no doubt but that it pays to use foundation in full sheets in the brood frames and sections, especially in the latter. If you think you cannot afford so much, you should at least have a strip for a starter. In order to get straight, even combs a starter is almost absolutely necessary. You can use a starter of any width from one inch up to full sheets. In sections by far the best results are obtained by using full foundation. The illustration shows the different styles of starters that may be used.

When using wax to fasten comb foundation, take a five pound lard pail with about three inches of wax and set it in a vessel of hot water on the stove. Place two wax tube fasteners in the melted wax long enough to heat thru thoroly, so they will not chill the melted wax and clog. Have the foundation ready and in place. Grasp the handle of the tube with the thumb, closing the air hole; place the tip of the tube where you want to start; raise the thumb to give air; and move along to stopping place and close the air vent. Then return the tube to the melted wax to keep warm and refill while using the other tube. With a little practice a neat, clean job can be rapidly done.

FIG. 18.—VAN DEUSEN WAX-TUBE FASTENER
A neat, clean job can be done with a little practice.

There are many other uses made of beeswax. It is used extensively in many of the arts, in making grafting wax, and floor wax, in ironing, sewing and in various other ways.

GENERAL NOTES

Missouri will support on an average one colony of bees to every ten acres and enable them to store close to thirty pounds of the best quality of surplus honey each year.

While a single bee has been enticed over five miles by scent and bait, it is rather an exceptional case. The average profitable pasture limit of a colony of bees will not extend more than a mile from the hive. On warm, still days the distance is greater, while in cool, cloudy, windy or unsettled weather it is less. If neighbors keep bees there may be overlapping of pasturage so as to reduce the number of colonies each may profitably keep.

The bees gather the nectar from flowers and make it into honey. The man who sells a hive that makes honey or guarantees success without any care is a fakir to say the least. There is no such thing as a non-swarming strain of bees and no hive or system of manipulation can possibly do more than help regulate swarming.

If beekeeping were a sure thing, it would be in a class by itself. Energy, skill, and brains, with a definite aim, bring success and pleasure in beekeeping.

Bees can be successfully cared for by a girl, boy, or woman with less capital and fewer accidents than in milking a gentle cow.

A first-class bee smoker and bee veil are indispensable to the apiarist. A sting on the face is needless as well as unpleasant, so the wearing of a bee-veil is classed as good common sense by the bravest.

FIG. 19.—THE RIGHT WAY   THE WRONG WAY
    Good shade   Exposure to hot sun
    No weeds crowding   Weeds crowding
    Level hive   Tilted hives
    Good hive   Box hives
    Protection from   No protection
      winter wind

Blow a little smoke into the entrance before opening hive, if you are timid or the bees show signs of being cross.

Do not handle your bees early or late in the day or in cool weather, as you will chill the brood and may later mistake the result for "foulbrood."

Select the warm part of the day for handling your bees when most of the old bees will be in the field, making it much easier for both you and the bees.

Keep the hives on bricks or stones; level side ways with front lower so water will not flow in. Four or six inches high is out of the way of the valuable toad. Keep the grass and weeds down so as not to obstruct the flight of your bees.

When you are stung, scrape the sting off at once before much poison is injected and then pass your hand over the smoker to remove the smell of the sting as it infuriates the other bees.

After being stung several times in the season, you will become immune to the swelling and will not mind it. An occasional "bite" will only keep up the immunity.

The bee is unjustly accused of many crimes which it does not commit. Birds and hornets injure fruit. The bees only try to save the juice of the injured fruit.

Read reliable works on bees, choosing those which stick to facts and are in plain language. If at first you are not successful, find out where you have erred; and adjust your work to suit the nature of the bees and you will soon be a successful beekeeper.

If the old box-hive or gum with deadly sulphur fumes is good enough for you and you see nothing in the "new-fangled" movable frame hive, you are not only a failure but a menace to beekeeping.

When you market your own honey ripened and sealed by the bees, it needs no pure food label.

Sell only pure first-class honey, using any "off in flavor" for feeding or for kitchen purposes.

Extracting frames with starters or full medium foundation is one of the best means to induce bees to store surplus honey.

Sixteen square inches of capped honey weighs about one pound. Capped honey when extracted weighs twelve pounds to the gallon.

All pure extracted honey will granulate or "sugar"—some very quickly. It is best to put it into one quart fruit jars at once and save future trouble.

To bring candied or sugared honey back to its former liquid state, put the jars in water and heat it slowly to about the scalding point (150°F.). Keep it at that temperature until all sugar is dissolved. It will then remain liquid for two or three months. Boiling honey spoils its flavor and color.

Candied honey in a fruit jar is a most delicious sweet and readily served.

Extracted honey which has been strained thru a cheesecloth to take out chips of comb is extracted honey and not "strained honey."

If the owner and the bees both get the "swarming fever," failure is sure.

Make haste slowly in the increase of colonies. Two strong colonies are enough to begin with and may need checking if they increase too fast.

If a swarm of bees is hived without a queen, a few bees will come out and run about as if searching for something valuable. This alarm is spread to the whole swarm and soon all are on the wing returning to the parent hive.

FALL NOTES

Remember to take surplus honey and no more. If a colony needs feeding do so before cold weather, so as to be able to cover it for winter.

Fruit juice is not a reliable winter food.

Store supers in a cool, dry room for winter and put them on early in the spring or the moths will destroy the combs.

Pile supers so that mice cannot get into them.

Protect all hives left out for the winter by wrapping them with tar paper and running wire around to keep it on, of course, leaving the entrance open.

Attend to the needs of your bees in time and let luck and superstition strictly alone.

To prepare bees for winter three things are essential: first, plenty of good wholesome honey; second, clean dry hive; third, proper protection to prevent sudden changes in temperature.

WINTER NOTES

Place the hives facing the south or east on well-drained ground with the hive about six inches above the ground and protect them as well as possible from the winter winds.

When bees are wintered in doors, the room should be kept at 40 degrees Fahrenheit, dark, dry, well ventilated, and free from noise or jar. Each hive should have 25 or 30 pounds of honey to prevent spring dwindling.

Leave two or three stands out as an index to pollen supply in the spring.

Do not attempt to keep your bees confined to the hive except in transportation.

Prepare your hives and other necessary things in winter.

Take a heavy piece of wire about two feet long and bend the end about two inches L-shape to rake out dead bees occasionally in winter.

SPRING NOTES

This is the dangerous season for the bees. Starvation may be overcome by feeding with the division board feeder.

Feed early in the spring before warm weather.

Keep rye flour in a box on some clean, bright hay in a sunny shed, out of the reach of stock, for the bees to work on every warm day until nature furnishes plenty of pollen, which they will collect in place of the rye flour. This will help to prevent spring dwinding.

It is the March, April, and May bees that are most valuable in producing strong colonies and abundant surplus of honey.

Do not raise drones. Cut out drone comb with the little blade of your pocket knife and make wax of it. A frame of drone comb in the brood chamber means a dollar lost.

The bees waste time and honey rearing drones which only eat up your profits.

The use of three-inch strips of comb foundation secures straight combs, helps the bees, and checks the over-production of drone cells. Anyone using comb foundation more than doubles the ability of the bees.

Send wax to a bee supply house and get foundation, hives, supers, and other necessary equipment in exchange for it.

SUMMER NOTES

The honey harvest is in short periods or flows of from a few days to a few weeks at best. If you have the bees in proper shape, they will get the honey when it is to be had. Keep your bees ready.

If you work for comb or chunk honey for home use, it will pay to use the deep super as it can also be used to the best advantage for extracted honey.

In case you should want some nice section honey, fill a super of sections with thin super foundation and place it under a half filled extracting super above the queen excluder. If the honey flow continues, you will get your reward.

You will get more than twice the honey after you have plenty of combs by extracting the honey and returning the empty combs for refilling.

When you have from four to six stands of bees, you need a two frame rapid reversible honey extractor and a solar wax-extractor.

Do not extract honey under any circumstances until it has been capped over by the bees, as all uncapped honey is "green" and will sour and spoil.

Early morning sunshine on the bees starts them off early; dense shade the rest of the day makes comfort, checks swarming, and prevents many bees getting overtaken by evening darkness.

When bees "hang out" or "lay out," the swarming fever, lack of room, too much heat, or some other fault needs a remedy at once. Do not neglect them unless you want to stand the loss. A hot hive in a hot place will drive away any swarm that can fly.

Raise your queens from the best colonies when making artificial swarms.

Give a queenless colony a frame of brood containing eggs from your best colony and in twenty-two to twenty-six days they will probably have a laying queen from your best colony. However, she may be ruined by mating with a drone from the poorest colony in the neighborhood. To prevent this, try to get your neighbors to requeen with good stock.

Early in the spring make a limited number of strong swarms as described under "How to Form Nuclei."

The old queen and the older bees go with the first swarm.

When a swarm is without brood, contented, and inactive, the bees should be shaken on a sheet in front of the hive and as the bees pass in the queen will generally be seen. Kill her at once and in a day or two give the bees a frame from which to raise a new queen. In case the old queen is not found within a reasonable length of time give them a couple of combs containing eggs from one of your best hives and in six or seven days they will have queen cells. If no queen cell appears on the two combs, they probably have a good-for-nothing queen that must be gotten out and the hive requeened.

Draw a couple of extra good frames from an over populous hive and exchange them for the poorest frames of a weaker and less prosperous colony and observe the effect. Leave all bees in their own hive.

When exchanging combs brush them clean to a bee.

Take the two outside frames from the weaker hive and put them in the center of the stronger and in exchange take two from the inner part of the stronger and put them in the center of the weaker colony.

Do not take chunk honey or extract honey from the brood chamber unless the bees get the swarming fever beyond endurance and make this necessary.

Expect no more honey from a swarming hive than eggs from a setting hen and there will be no disappointment.

All bees become robbers if tempted with exposed sweets in time of scarcity.

Do not smear. Learn to work with care. You will be pleased and the care of bees will not be drudgery. There are times when the honey flow ceases all at once and leaves the bees with nothing to do but protect their stores. Even the most gentle bees are cross, irritable and furious if disturbed at such times. It is folly to try to handle them.

All bees are gentle in a copious honey flow and exactly the opposite when the supply is suddenly cut off.

Bees should not be kept near stock of any kind. The orchard is a good place, also the yard if shaded, or a grove will answer the purpose very well.

When bees swarm, do not ring bells, beat on tin pans, shout, or do other foolish things.

HONEY PLANTS

The bee collects nectar and pollen from a great many different kinds of flowers, tho most of the honey comes from a very few. The flower must be one not too deep for the bee to reach the bottom and it must have enough nectar to attract the bee. The honey bee cannot profitably collect nectar from red clover because its tongue is too short. The bumblebee with a long tongue is especially adapted for the red clover.

Beginning in the spring we have maple blossoms and then the fruit blossoms which provide some nectar and pollen; later come the locust, lindens and the all important—the real Missouri honey plant—white or wild clover. A very large part of the honey in this state comes from white clover. Bee clover or common sweet clover is rapidly coming to the front as a honey plant and a soil builder for waste places. Later we have the summer and fall blossoms, such as alfalfa, golden rod, smart weeds, buckwheat, Spanish-needle and other similar blossoms. The late fall blossoms usually produce an inferior grade of honey. For a more complete list of honey plants one should refer to a good text-book.

ROBBING

Some types of bees are worse than others about robbing, but their untiring industry and instinct for hoarding make all bees by nature capable of becoming robbers if tempting sweets are left unprotected. It is always well to bear this in mind. Be careful that you do not smear honey about the hives and ground. Also keep all colonies strong, for a dwindling, queenless colony furnishes temptation for robbers.

The beekeeper's best motto is "keep all colonies strong." The robber is merciless, taking from the weaker colonies until they swarm out even in the fall or early spring when there is no honey left. Robbing is the cause of many freak actions of bees.

Robber bees are sneaking in action and become stripped of so many hairs as even to look smooth. They will tear down combs, dropping sawdust-like chips on the bottom of the hive in their hurry to make away with the ill-gotten gain.

Should your bees get to robbing, contract the entrance of the hive being robbed to a working limit, of about one inch. Coarse hay or weeds put over the entrance, as in moving bees, will generally stop it quickly. A robber will not go where the guards can get hold of it, so you are taking advantage of a weak point. Sprinkling with cold water also adds to their fear.

ENEMIES OF BEES

The greatest enemy of the bee is the almost criminal neglect and abuse of man. Most other enemies are easily overcome by the bees themselves when properly cared for. The moth may cause trouble in a gum or box-hive, or even in a regular hive if the colony is weak and dwindling, but it is soon controlled where the colony is kept strong with a young, vigorous queen. The toad, bee-bird, and other similar enemies, may be kept from bees with a little care.

LITERATURE

The beekeeper who desires more detailed information concerning the honey bee and beekeeping is referred to the various technical reports, text-books, and bee journals which treat the subject in a more exhaustive manner. There are too many of these to be listed here, but anyone desiring a list of the particular publications which will be of greatest value to him may secure this on request from the Missouri Agricultural Experiment Station if he will include a brief description of what he is doing in the beekeeping line.

BEE SUPPLIES AND QUEENS

There are a number of responsible men and firms in the state that handle bee supplies and queens. As complete a list of these as it has been possible to collect to date will be sent on request by the Agricultural Experiment Station. Each request should be accompanied with a statement of the number of colonies, kind and amount of honey produced, most important native honey plants, and any other local information regarding beekeeping which may be available.

www.ingramcontent.com/pod-product-compliance
Lightning Source LLC
Chambersburg PA
CBHW062233220526
45471CB00009B/3467